A Beginners Guide to the Mechanics of Wrist and Pocket Watches

Including the History of Their Development and Some Famous Watch Makers

Mechanical Details of Watches — Divisions in the Watchmaking Trade — Watch Chain — Hair Spring — Ticks of a Watch — Watch Glasses — A Good Watchmaker — Swiss Watches — Watches Imported into England — Duties on Watches — Depression of Watch Trade — Frauds in Watches — Epitaphs on Watchmakers — Moore's Lines with a Watch — Proverbs on Watches — Sheridan's Strategem to get a Watch — Washington's Secretary's Watch — Anecdote of a Lord Mayor of London — Un Poisson d'Avril — How a Clergyman got a Watch from a Thief — Watch with Lettered Dial — How a Watchmaker got Trade — Watch Papers — Poetry of — Watch Stands.

We here propose to give a short account of some of the curiosities of the mechanical details of watches. A good three-quarter plate watch as usually made requires no fewer than one hundred and thirty-eight distinct pieces in its frame, train, escapement, potence, fusees, arbors, clicks, ratchets, and other nicely-contrived and adjusted constituents. To these appliances must be added the chain, which contains sixty-three links and forty-two rivets to every inch, and being generally six inches in length comprises six hundred and thirty pieces; thus swelling the contents of a common detached lever-watch to seven hundred and sixty-eight separate pieces, to construct which gives occupation to no less than thirty-eight or forty different kinds of artificers. Babbage, in his 'Economy of Manufactures,' tells us that the division of labour cannot be successfully practised unless there exist a great demand for its produce; and it requires a large capital to be employed in those arts in which it is used. In watchmaking it has been carried, perhaps, to the greatest extent. It was stated in evidence before a Committee of the House of Commons, that there are a hundred and two distinct branches of this art, to each of which a boy may be put apprentice; and that he learns only his master's department, and is unable after his apprenticeship has expired, without subsequent instruction, to work at any other branch. The watch-finisher, whose business is to put together the scattered parts, is the only one out of the hundred and two persons who can work in any other department than his own.

The chain used in connecting the main spring and fusee in watches and clocks is composed of small pieces of sheet-steel, and it is of great importance that each of these pieces should be of exactly the same size. The links are of two sorts; one of them consisting of a single oblong piece of steel with two holes in it, and the other formed by connecting two of the same pieces of steel, placed parallel to each other, and at a small distance apart, by two rivets. The two kinds of links occur alternately in the chain; each end of the single pieces being placed between the ends of two others, and connected with them by a rivet passing through all three. If the rivet-holes in the pieces for the double links are not precisely at equal distances, the chain will not be straight, and will consequently be unfit for its purpose.

The hair-spring of a watch is no less remarkable for the extreme delicacy of its construction than for the proof which it gives of the great value that can be given to a small piece of steel by manual labour. Four thousand hair-springs scarcely weigh more than a single ounce, but cost often more than 1,000*l.* "The chisel of the sculptor," says Mr. Thompson, "may add immense value to a block of marble, and the cameo may become of great price for the labour bestowed; but art offers no example wherein the cost of the material is so enhanced by human skill as in the balance-spring." The pendulum-spring of a watch, which governs the vibrations of the balance, costs at the retail price twopence, and weighs fifteen one-hundredths of a grain, while the retail price of a pound of the best iron, the raw material out of which fifty thousand such springs are made, is the same sum of twopence.

A watch it is said ticks seventeen thousand one hundred and sixty times in an hour, that is four hundred and eleven thousand eight hundred and forty times in a day, and one hundred and fifty millions four hundred and twenty-four thousand five hundred and ten times in a year, allowing the year to be three hundred and sixty-five days and six hours long. If a watch went, as some will, for one hundred years, it would beat fifteen thousand and forty-two millions four hundred and fifty-six thousand times.

Watch glasses were formerly made in England by workmen who purchased from the glass-house globes of five or six inches in diameter, out of which by means of a piece of red hot tobacco pipe, guided round a pattern watch-glass placed on the globe, they cracked five others; these were afterwards ground and smoothed on the edges. In the Tyrol the rough watch-glasses are supplied at once from the glass-house; the workman applying a thick ring of cold glass to each globe as soon as it is blown causes a piece of the size of a watch-glass to be cracked out. The remaining portion of the globe is immediately broken and returned to the melting pot. Formerly this process could not be adopted in England with the same economy, because the whole of the glass taken out of the pot was subject to excise duty. In

England at the present time, the crown glass, used for enclosing the dials of clocks and watches, before being moulded or bent into the required form, is first cut into circular shape by means of a circle-cutter, which consists of a circular board, covered with wash-leather, which is made to revolve on a pivot by one hand of the operator, while with the other hand he presses down a diamond on to the glass; the diamond is fixed at the end of an adjustable arm, which traverses a slot, the exact diameter of the circular plate to be cut being regulated by an index fixed at the side of the slot. The circular flat plates, which are removed to moulds turned out of solid firestone, the sinking of the moulds being of flat elliptical section, are put into one or other of the furnaces, according to the size of the glass to be bent; while in the furnace, the mould is kept in continual circular motion by the long iron rod of the operator, until the glass sinks into the required form. The grinding and polishing the edge of the glass is the next operation. By a simple and ingenious contrivance the edges of clock and watch glasses are ground. The operator stands in front of the work, with his right hand turning round a handle, placed vertically above the bench, and with his left hand holding a sort of hood, or cup, supplied with emery powder, by which the grinding is partially effected; the glass to be ground is temporarily fixed on a boxwood mallet, by means of cement, at the top of a spindle, which passes through the bench, and to which rapid motion is communicated by a round band from a horizontal wheel below the bench, turned by the right hand of the operator. After the grinding is completed, the edge of the glass is smoothed with pumice-stone, and finally polished with oxide of tin, usually called putty-powder.

"To become a good watchmaker," says Berthoud, "it is necessary to be an arithmetician in order to find accurately the revolutions of each wheel; a geometrician, to determine correctly the curve of the teeth; a mechanician, to find precisely the forces that must be applied; and an artist, to be able to put into perfect execution the principles and rules which these sciences prescribe. He must know how fluids resist bodies in motion, and be well acquainted with the effects of heat and cold in different metals; in addition to these acquirements he must be endowed by nature with a happy genius to be able to apply them all in the construction of an accurate measurer of time."

As to Swiss watches. It has been stated that the beginning of the Geneva watch-trade was owing to the bloody persecution of the Protestants in France towards the close of the sixteenth century, which caused some French watchmakers to take refuge in Switzerland. In 1587 Charles Cusin, of Autun, in Burgundy, settled in Geneva as a manufacturer of watches, which were then sold for their weight in gold. He had many scholars, and his success naturally drew labour from less profitable employment, and spread the watchmaking trade very rapidly. But the substantial introduction of watch manufacture into Switzerland, which, by the extensive employment of female labour, competes successfully with England, occurred in this manner:—In 1679 a native of Switzerland, after a visit to this country, took home with him from London a watch, which was a novelty in his mountain land, and greatly excited the curiosity and interest of all who saw it there. On its getting out of order the owner of the watch trusted it for repair to a mechanic, named Daniel John Richard, of La Sange, who, after attentively examining the mechanism of it, determined to construct a watch after its model. He according spent a year in making the requisite tools, and in six months afterwards completed a watch. He next set up in the trade of a watchmaker, and supplied his neighbourhood with watches. He died in 1741, leaving five sons, who pursued their father's profitable business. The trade has been largely developed during the present century, and Swiss watches are in universal demand on account of their cheapness. Mr. J. Bennett, whose advocacy of female labour in the watchmaking art has rendered him very obnoxious to some persons, calculated a few years ago that no less than two hundred thousand Swiss watches were annually imported or smuggled into England, while the whole produce of the English watchmakers did not exceed one hundred and eighty-six thousand. In the valleys and on the slopes of the Jura mountains, during a winter which lasts six or seven months, the inhabitants, who are kept indoors by the inclemency of the weather, vigorously pursue some one or other of the several departments of the watch-trade. Mr. Bennett states that on a visit paid by him to Neufchâtel, where the largest number of the Swiss watches are made, he found no less than twenty thousand women employed in making the more delicate parts of the watch movement, and earning upon an average fifteen shillings per week each. They produced annually about one million five hundred thousand watches, besides movements, for the American markets.

An old astronomical watch of Swiss manufacture was exhibited by Mr. Morgan to the Society of Antiquaries in 1850. It was made at Geneva by Jean Baptist Duboule, and bore a striking contrast to the Geneva watches now in use. It was of large dimensions, being three inches and a half in diameter and two inches thick, and was in fact a clock-watch; it had an alarum, and showed on the different circles of its face the hours and the period of the day, whether morning, noon, evening, or night, by certain allegorical figures engraved on a revolving disc, which changed suddenly, and presented themselves at proper times. The days of the week were also represented by revolving figures, the days of the month, the name of the month, and the number of days it contained, together with the sign of the zodiac in which the sun was, the age of the moon and its phases, and the four seasons of the year, "as they in their circle run" by allegorical figures engraved on a revolving disc in the same manner as the four periods of the day. This complicated and beautiful piece of mechanism was not quite in its original state; some former possessor about a century and a half ago caused a new balance-wheel with pendulum-spring to be added, to make it perform more regularly. It was enclosed in a travelling-case of black leather, ornamented with silver studs.

The importation of cheap foreign watches in great numbers into England obviously must have a depressing influence upon the English trade, by lessening the demand, as numbers of unwary persons who study a false economy, purchase those of Swiss make, on account of their tempting cheapness. The number of watches imported from Switzerland into this country in 1858 was three hundred and forty-six thousand eight hundred and ninety-four. The returns of Goldsmiths' Hall afford the following statistics of English manufacture in the number of watch-cases hall-marked, in 1858:—London, eighty-three thousand six hundred and fourteen silver, and twenty-six thousand eight hundred and seventy gold; in Coventry, sixty thousand silver. By our transatlantic cousins British watches were preferred; thus, in 1857 we exported to America fourteen thousand one hundred and forty-one watches of home manufacture, and four hundred of foreign make. On the contrary, Australia showed a preference for foreign watches, by taking six thousand seven hundred and twenty-two of them, and only three thousand and eighty-two of British manufacture. During the year 1860 one hundred thousand foreign watches passed through the custom-house for home use. The number of foreign watches imported into England free of duty during the first seven months of the year 1865 was eighty-six thousand one hundred and fourteen, and during the first eight months ninety-six thousand four hundred and one as against eighty-seven thousand two hundred and eighty-two in 1864, and one hundred thousand nine hundred and thirty-eight in 1863, periods corresponding with the latter months.

The parliamentary inquiries which have been instituted from time to time into various matters connected with the clock and watch trade have served to enlighten us as to its extent, and the regulations by which it was controlled. All the clockmakers and other persons using their trade within London and ten miles therefrom were incorporated into one body, with powers to make by-laws for the government of those persons who used the trade throughout England, and to control the importation of foreign clocks and watches into this country, and to mark such as were imported. By the custom of marking foreign watches with the stamp of the Clockmakers' Company, of marking each English watch with the maker's name, and of the stamping of the cases at Goldsmiths' Hall, the number of watches produced in England became tolerably well ascertained.

In a report from a committee of the House of Commons in 1818 it is stated that the number of gold and silver watch-cases marked at Goldsmiths' Hall in the year 1796 amounted to one hundred and ninety-one thousand six hundred and seventy-eight; and in the year 1798, when duties were imposed on clocks and watches, and a licence to deal therein was made necessary, the number of watch-cases hall-marked was reduced to one hundred and twenty-eight thousand seven hundred and ninety-eight, showing a decrease of sixty-two thousand eight hundred and eighty. Although these duties were repealed, the manufacture never recovered itself. In 1808 the number of watch-cases marked was most approximate, it being one hundred and eighty thousand three hundred and eighty-nine. From this year the number diminished, so that in 1815 only one hundred and thirty-four thousand two hundred and sixty-nine watches were hall-marked; and in 1816 only one hundred and two thousand one hundred and twelve, rather more than one-half of the number marked in 1796. Babbage, in his 'Economy of Manufactures,' says that the number of watches manufactured for home consumption was in 1798 about fifty-thousand. If this supply was for Great Britain only, it was consumed by about ten and a half millions of persons.

Two causes are clearly assignable for the signal declension in the manufacture of watches, as mentioned above, namely, the imposition of a duty on clocks and watches in 1797, and the illicit importation of foreign watches. During the Pitt administration, an act was passed levying certain duties upon clocks and watches, to be paid by the possessors of them. It was ordered by this act that, after July 5th, 1797, an annual duty of five shillings should be charged on every clock or timekeeper; for every gold watch kept, worn, or used, ten shillings; and for every silver or metal watch, two shillings and sixpence per annum. From this pernicious tax the Royal Family were exempt, as were also hospitals and churches, farm-servants, soldiers, and sailors; nor did it extend to the stock of pawnbrokers and clock and watch makers; but it required them to take out annually, at a cost of two shillings and sixpence, a licence to sell those goods. The duty in a single year amounted to 48,820*l*. But the act was repealed in 1798, owing, it is said, to the representations of a committee of watchmakers convened in Clerkenwell, who proved to Mr. Pitt that a piece of work valued at 500*l*. had been manufactured out of materials which did not in their rough original state cost so much as sixpence. Dr. Rees says, "This representation, we are credibly informed, induced the great statesman to abandon his plan of taxing an article, the value of which depended so much upon the ingenuity and labour by which thousands were entirely supported; nay, further, on learning that the French and Swiss could afford to sell three gold watches for the price of one English one, the same minister took off the duty of 16*s*. per ounce from watch-cases of that metal, and substituted 1*s*., the price of marking at Goldsmiths' Hall." In the years 1797 and 1798, the Clockmakers' Company appealed to Parliament respecting the severity of the act. The petition was presented to the House on February 23rd, in the latter year, by Alderman Lushington, M.P., and after being read was referred to a Committee, with other petitions on the same subject. On the 27th the bill for repealing the duty on gold and silver watch-cases was read a third time, and passed up to the House of Lords. On March 14th, the Chancellor of the

Exchequer stated in the House of Commons that the repeal of the watch act would cause a deficiency of 200,000*l.*, and as this tax was proved by experience to press heavily upon a very industrious and useful part of the community, he would, in order to meet the rules of the House, simply propose to repeal the tax on clocks, watches, and time-keepers; and on a future day bring forward another measure to meet such deficiency. His motion formally was, "that the tax on clocks, watches, and timekeepers made last session should cease and determine," which was agreed to. On the following day Mr. Hobart brought up a report of this resolution of the Committee of the whole House, which was read and assented to. The act imposing the duty had occasioned great distress among the watchmakers of Clerkenwell.

On June 6th, 1814, Sir William Curtis presented a petition to Parliament from the clock and watch makers, stating that they exported goods to the value of 500,000*l.* annually; and that their trade was in danger of being ruined by the importation of foreign clocks and watches to which the names of English workmen were affixed. In the years 1816 and 1817, great distress prevailed among the watchmakers of Clerkenwell, many of whom were reduced to the extremity of want. At no period since 1797 had there been such a lack of employment; the earnings of such as obtained work in the last three months of 1816 averaged seven shillings and sixpence per week. A philanthropic society was formed, and upwards of three thousand indigent workmen were relieved from its funds. Two pawnbrokers in Clerkenwell alone had watchmakers' tools in pledge to the amount of 190*l.*, a sum not exceeding one-third of their real value. This depression in the clock and watch trade was attributed to the extensive illicit importation of such articles from Geneva and other parts of Switzerland, the movements being brought over into this country, put into English cases, and then sold as English watches; while others were made up complete so as to resemble those of British manufacture. Babbage, writing in 1835, says that in the watch trade the practice of deceit, in forging the marks and names of respectable makers, had been carried to a great extent both by foreigners and natives; and the effect upon our export trade had been most injurious, as the following extract from the evidence before a Committee of the House of Commons proved:—

Question. How long have you been in the trade?

Answer. Nearly thirty years.

Q. The trade is at present much distressed?

A. Yes, sadly.

Q. What is your opinion of the cause of that distress?

A. I think it is owing to a number of watches that have been made so exceedingly bad that they will hardly look at them in the foreign markets; all with a handsome outside show, and the works hardly fit for anything.

Q. Do you mean to say, that all the watches made in this country are of that description?

A. No; only a number which are made up by some of the Jews, and other low manufacturers. I recollect something of the sort years ago, of a fall off of the East India work, owing to there being a number of handsome-looking watches sent out, for instance, with hands on and figures, as if they showed seconds, and had not any work regular to show the seconds: the hand went round, but it was not regular.

Q. They had no perfect movements?

A. No, they had not; that was a long time since, and we had not any East India work for a long time afterwards.

In the home market inferior but showy watches were made at a cheap rate, which were not warranted by the maker to go above half-an-hour; about the time occupied by the Jew pedlar in deluding his country customer.

According to the census of 1861 there were eight hundred and seventy-seven watch and clock makers in Clerkenwell.

At Lydford, in Devonshire, is the following epitaph on a watchmaker:—

"Here lies in a *horizontal* position,
the outside *case* of
George Routleigh, watchmaker,
whose abilities in that line were an honour to his profession.
Integrity was the *mainspring*, and prudence the *regulator*
of all the *actions* of his life;
Humane, generous, and liberal, his *hand* never *stopped*
till he had relieved distress:
So nicely *regulated* was his *movements*,
that he never *went wrong*,
except when *set a-going*
by people who did not know his *key*:
Even then he was easily *set right* again.
He had the art of disposing of his *time*
so well,
That his *hours* glided away in one
continual *round* of pleasure and delight,
Till an unlucky *moment* put a *period* to his existence.
He departed this life November 14, 1802,:
Aged 57, *wound up*,
in hopes of being taken in *hand* by his *Maker*:
And of being thoroughly *cleaned*, *repaired*, and *set a-going*
for the world to come."

In Uttoxeter churchyard is the following epitaph,

which was written on himself by a watchmaker, who was fond of good ale:—

"Here lies one who strove to equal time!
A task too hard, each power too sublime.
Time stopt his motion, o'erthrew his balance-wheel;
Wore off his pivots, though made of harden'd steel;
Broke all his springs, the verge of life decay'd,
And now he is as though he'd ne'er been made.
Not for the want of oiling—that he tried,
If that had done—why then he ne'er had died."

At Shrewsbury is the following epitaph on a watchmaker:—

"Thy movements, Isaac, kept in play,
Thy wheels of life felt no decay
For fifty years at least;
Till by some sudden secret stroke,
The balance or the mainspring broke,
And all the movements ceas'd."

Moore sent the following lines, "To a Boy with a Watch:"—

"Is it not sweet, beloved youth!
To rove through Erudition's bowers,
And cull the golden fruits of Truth,
And gather Fancy's golden flowers?

"And is it not more sweet than this,
To feel thy parents' hearts approving,
And pay them back in sums of bliss,
The dear, the endless debt of loving?

"It must be so to thee, my youth;
With this idea toil is lighter;
This sweetens all the fruits of Truth,
And makes the flowers of Fancy brighter.

"The little gift we send thee, boy,
May sometimes teach thy soul to ponder,
If indolence or syren joy
Should ever tempt thy soul to wander.

"'Twill tell thee that the wingèd day
Can ne'er be chain'd by man's endeavour;
That life and time shall fade away,
While heaven and virtue bloom for ever."

The following two old proverbs relate to watches:—"You may be a wise man though you cannot make a watch;" "A man, like a watch, is to be valued for his goings."

This story is told of Sheridan's stratagem to get a watch:—Harris, the proprietor of Covent Garden Theatre, who had a great regard for Sheridan, had at different times frequent occasion to meet him on business, and made appointment after appointment with him, not one of which was ever kept by Sheridan, who during his life wasted much of old Time's sand. At length Harris, wearied out, begged his friend Palmer, of Bath, to see Sheridan, and tell him that unless he kept the next appointment made for their meeting, all acquaintance between them must end for ever. Sheridan expressed great sorrow for what in fact had been inevitable, and positively fixed one o'clock the next day to call upon Harris at the theatre. At about three o'clock he made his appearance in Hart Street, where he met Tregent, the celebrated French watchmaker, who was extremely theatrical, and had been the intimate friend of Garrick. Sheridan told him that he was on his way to call upon Harris. "I have just left him," said Tregent, "in a violent passion, having waited for you ever since one o'clock." "What have you been doing at the theatre?" said Sheridan. "Why," replied Tregent, "Harris is going to make Bate Dudley a present of a gold watch, and I have taken him half-a-dozen that he may choose one for that purpose." "Indeed," said Sheridan. They wished each other good day and parted. Sheridan proceeded to Harris's room, and when he addressed him it was evident that his want of punctuality had produced the effect which Tregent had described. "Well, sir," said Harris, "I have waited at least two hours for you again; I had almost given you up, and if—" "Stop, my dear Harris," said Sheridan, interrupting him, "I assure you these things occur more from my misfortunes than my fault; I declare I thought it was but one o'clock, for it so happens that I have no watch, and to tell you the truth, am too poor to buy one; but when the day comes that I can, you will see I shall be as punctual as any other man." "Well, then," said the unsuspecting Harris, "if that be all, you shall not long want a watch, for here" (opening his drawer) "are half-a-dozen of Tregent's best; choose any one you like, and do me the favour of accepting it." Sheridan affected the greatest surprise at the appearance of the watches; but did as he was bid, and selected certainly not the worst for his present.

When Washington's secretary excused himself for the lateness of his attendance, and laid the blame upon his watch, his master quietly said, "Then you must get another watch, or I another secretary."

The following original anecdote has been communicated to the author by Mr. John Bullock, of Sevenoaks:—A gentleman, who now fills a responsible situation in one of the national exhibitions, some years since became bail for a friend; and went before the then Lord Mayor of London to justify, when this conversation took place:—

Lord Mayor. What is your name?
Bail. W—— F——.

Lord Mayor. Well, Mr. F——, you have come here to be bail for your friend; what are you?

Bail. An articulator, my lord. (An articulator is one who puts in correct order the bones or skeletons of animals.)

Lord Mayor's Clerk. What is that?

Lord Mayor. Oh! I know, you have to do with clocks and watches.

The Bail, being unwilling to expose my Lord, bowed assent; and so the chief magistrate of the City passed for wise, and matters went off satisfactorily.

It is related that a French lady having stolen a watch from a friend's house on the first of April, endeavoured after detection to pass off the affair as *un poisson d'Avril*, an April joke. On denying that the watch was in her possession, a messenger was sent to her apartments, where it was found upon a chimney-piece. "Yes," said the thief, "I think I have made the messenger a fine *poisson d'Avril.*" But the magistrate before whom she was taken said she must be imprisoned until the first of April in the ensuing year, *comme un poisson d'Avril.*

Dr. Bigsby tells the author that an old friend of his, a Nottinghamshire clergyman, being in London, went one evening to the pit of Drury Lane Theatre to see some famous actor of the day—Kemble or Kean. There was a great rush for admission at the entrance, and, while sorely pressed by the crowd, he felt somebody's hand busy with his watch-pocket. Searching to see if his watch was safe, he found that his fob was empty; whereupon, remarking a suspicious-looking fellow immediately in advance of him, he promptly charged him, but in a whisper only, with the robbery. "You've got my watch," said he. At the next moment a watch was slipped into his hand by the party addressed, with the words, "Say nothing about it; here it is." The clergyman conveyed the article to a more secure place of deposit, and thought little more of the matter until his arrival at his inn, after the close of the performance. On going up to his bedroom at the 'Bull and Mouth,' the first object he saw was the black ribbon, with the seals appendant, of his own watch, hanging at the bed's head; and on examining his coat-pocket, he found a most magnificent watch, of the value of some forty or fifty pounds, that had doubtless been stolen a short time before it had been handed to himself by the thief in question. He repeatedly advertised the watch in the newspapers, but it was never claimed; and he used afterwards to exhibit it to his friends, and tell its story to them, with a comic expression of satisfaction at the idea of having fleeced an Arab of his spoil in so unconscious a manner.

Apropos of watch stealing, we are reminded of a silver hunting-watch that belonged to one of the whippers-in of Lord Middleton's fox-hounds in Nottinghamshire. It had the words "Jack Stevens" on its face instead of the usual figures, the letters representing eleven figures, and in lieu of the twelfth figure was a fox's head. Such an arrangement might, in many instances, save a watch from being stolen.

The following story is told of a poor watchmaker's trick to obtain employment:—He settled at a populous country town, where he was unknown, and had no trade. He contrived when the church door was opened daily, to send up his son to the church-tower unseen, to alter the clock. This the lad was enabled to do; and as every one swore by the church-clock, all the watches in the neighbourhood were repeatedly found to be wrong.

Consequently the owners of them sent them to the new-comer to be repaired and set right.

By the way we may add a few particulars relating to watch-papers as they were called, but which were as often of silk, velvet, and muslin, printed or worked with the needle, as of the material whence they derive their name. These papers were used in the outer cases of the large old-fashioned watches, before the introduction of the present compact form of such instruments; and were decorated with verses or devices, as tokens of love or friendship. The papers were frequently very neatly cut with elaborate designs, sometimes spreading over the whole field, while at others a circle or oval was left in the centre, on which was painted a miniature. One old paper had these lines engraved upon it:—

"Content thy selfe withe thyne estat,
And sende no poore wight from thy gate;
For why, this councell I thee give,
To learne to dye, and dye to lyve."

In another, the following lines were so regulated as to be printed in a circle without a break:—

"Onward—
Perpetually moving—
These faithful hands are proving
How soft the hours steal by:
This monitory pulse-like beating,
Is often times methinks repeating,
'Swift, swift, the hours do fly!'
Ready, be ready! perhaps before
These hands have made
One revolution more,
Life's spring is snapt,—
You die!"

Another version of these lines was set to music early in the present century:—

"Onward perpetually moving,
These faithful hands are ever proving
How quick the hours fly by;
This monitory pulse-like beating
Is often times methinks repeating,
'Swift! swift! the moments fly.
Reader, be ready, for perhaps before
These hands have made one revolution more,
Life's spring is snapt—you die!"

In an old silver-watch were these printed lines:—

"Time is—the present moment well employ;
Time was—is past—thou canst not it enjoy;
Time future—is not, and may never be;
Time present—is the only time for thee."

We find the following "Lines on a Watch-paper" in the 'Gentleman's Magazine' for May, 1797, with the initials M. E. L.:—

"Life's morning hours unthought-of fly,
Without watch;
Life's noontide hours neglected die
By deaf watch;
Life's eve's repentance makes us cry,
For stop watch;
Life's midnight hour then beats by sigh,
A death watch."

A watchmaker named Adams, who practised his craft early in this century at Church Street, Hackney, was fond of putting scraps of poetry in the outer cases of watches sent to him for repair. One of his effusions was as follows:—

"To-morrow! yes, to-morrow! you'll repent
A train of years in vice and folly spent.
To-morrow comes—no penitential sorrow
Appears therein, for still it is to-morrow.
At length to-morrow such a habit gains,
That you'll forget the time that Heaven ordains,
And you'll believe that day too soon will be
When more to-morrows you're denied to see."

Two watch-papers, executed at the commencement

of the reign of George III., the one of white cambric wrought in gold thread with the letters G. M. C. within a double circle of loop-chain work; and the other of white muslin, with the initials S. G. in brown hair, are still preserved. Printed watch-papers with the head of William, Duke of Cumberland, were published in 1746, and some of white and pink satin with the portrait of Queen Caroline were common about the year 1821. Perhaps the most famous watch-papers were those printed on the ice during the frost fair of 1814.

It seems that watchstands came into use about the middle of the seventeenth century, but their history is very uncertain. One of that date is still preserved; it is of carved oak, eight inches high, and in design much resembles some of the mural monuments of the period. Two scrolls form a sort of pediment above, with an escalop-shell in front. It has side columns with four large flowers surrounding the circular opening for the face of the watch, and the lower part is pointed. At the back is a curved channel to contain the chain, which when placed in the stand together with the watch was shut in with a sliding backboard. Immediately behind the shell is a round hole to admit a hook or nail, by which the stand was suspended against a wall. There is also extant another watchstand of carved wood, nine inches and three-quarters high, of German workmanship, of the time of Francis I., 1741–1765. It represents the two-headed eagle, the imperial crown being placed between the necks; a branch of laurel proceeds from the crown and falls on each side of the circular opening in the breast of the bird. The sockets of the wings are still visible, but the wings themselves are gone. The bird stands upon a base of three stages, and the whole is painted in various colours.

Some of the old watchstands represented Time, from whose finger the watch depended; others were in the shape of horses, camels, and elephants, with panniers, vases, and temples on their backs; and others of a more common description were in the shape of churches and castles, with the watch-face appearing in the turret and over the gateway.

During the last century ladies exercised their ingenuity in making watchstands of rockwork formed of bits of spa, galena, and coloured glass, garnished with green moss; or of shells and seaweeds; or of cardboard besprinkled with glittering frostings, and decorated with cut, folded, and twisted paper, and gold and silver foil and spangles.

CPSIA information can be obtained at www.ICGtesting.com
Printed in the USA
LVOW101150041212

309974LV00001B/381/P